STRIP THE EXPERTS

About the author

While doing this PhD in theoretical physics at Sydney University, Brian Martin started examining research papers arguing about the effect of Concorde exhaust on ozone. He became interested in the way different scientists came up with startlingly different conclusions. He wrote *The Bias of Science*, a detailed study of the ways scientists push their arguments one way or the other.

In 1976 he moved to Canberra to work at the Australian National University as an applied mathematician, and became involved in the debate over nuclear power. The pronuclear experts were using their titles and scientific authority to powerful effect, and he had lots of practice in developing ways to deflate them and their arguments.

He also became active in exposing and opposing attempts to suppress dissident viewpoints. He is co-editor of the book *Intellectual Suppression*.

In 1986 he moved to the University of Wollongong, officially becoming a social scientist. He has studied controversies over fluoridation, nuclear winter, pesticides and (with Gabriele Bammer) repetition strain injury.

Brian Martin has long been involved in the radical science movement, the environmental movement and the peace movement.

STRIP THE EXPERTS

Brian Martin

FREEDOM PRESS
London
1991

Published by
FREEDOM PRESS
84b Whitechapel High Street
London E1 7QX
1991

© Brian Martin & Freedom Press

ISBN 0 900384 63 8

Printed in Great Britain by Aldgate Press, London E1 7QX

Contents

Introduction *5*

1 Challenge the facts *11*

2 Challenge assumptions *21*

3 Discredit experts *33*

4 Discredit expertise *54*

Tips on dealing with the experts *65*

References *67*

Acknowledgements

Much of what I've learned about experts comes from involvement in campaigns over nuclear power and other issues; both supporters and opponents have provided me with many valuable lessons, intended or otherwise. I thank also those individuals, too numerous to mention, who have discussed experts with me over the years. Gabriele Bammer, Mark Diesendorf, Denis Pym, Vernon Richards and Sheila Slaughter gave me helpful comments on earlier drafts. Thanks to David Mercer for suggesting the title.

Introduction

How often have you found the experts lined up against you? It happens all the time. "Don't eat eggs — there's too much cholesterol." "Nothing can travel faster than the speed of light." "Smoking causes lung cancer." "There's no danger from these radioactive emissions."

In modern society, scientific experts are the new priests. They pronounce on all manner of things with the ultimate authority: scientific knowledge. To challenge the experts is heresy.

Yet it can be done. The experts are vulnerable in a variety of ways. You can dispute their facts. You can challenge the assumptions underlying their facts. You can undermine their credibility. And you can discredit the value of expertise generally. Their weaknesses can be probed and relentlessly exploited.

This booklet is designed for people who oppose a gang of scientific experts and want to strip them naked. It describes various methods you can use to do this, with examples such as nuclear power, fluoridation, creation science, smoking and health, and nuclear winter. Each of these shows how a small number of critics can mount remarkably effective challenges to a powerful scientific establishment.

Now, in some of these cases I actually agree with the experts. I've had to force myself to describe how the tobacco industry has challenged the experts on smoking and health. This is a useful exercise, because there is probably more to be learned when it's our favourite cause being attacked.

There are lots of other experts that I don't have time to discuss: the legal profession, investment advisors, traffic engineers, agricultural scientists, etc. Indeed, it's difficult to do most anything without running into some expert or other.

I've been asked, "But surely you don't mean to encourage challenges to *every* expert?" Why not? The experts have all the advantages: degrees, status, salaries, connections, positions. If they can't defend themselves against challenges, perhaps they should retire to safer occupations. In my opinion, the more open debate, the better.

However, most people only reject some experts — the ones they disagree with. When *their* experts are the authorities, they're happy enough.

This is a risky approach. Give some experts credibility and power, and it won't be easy to get rid of them — *any* of them. This applies especially to experts in politics and economics, whether they're defending socialism or capitalism.

Every powerful group — government, corporation, profession or church — has its own group of experts at hand to provide justifications for its power, privilege and wealth. What has happened is that most experts today are servants of power.

More than a century ago, the famous anarchist Michael Bakunin warned of the dangers of government by the experts. Anarchists oppose any system — including government — in which a small number of people dominate over others. Instead, decisions should be made directly by the people, on the basis of free and open dialogue. Knowledge is important, but it should be knowledge accessible and useful to the people. At the moment, much "expertise" is knowledge that is so specialised and esoteric that it is only useful to experts and their patrons. So, perhaps, attacks on expertise are warranted today as part of efforts to make knowledge more relevant to people. In a more egalitarian, participatory society, knowledge would be prized. But it would be knowledge at the service of all, not knowledge for elites.

Pardon the lecture. You don't have to be an anarchist to be critical of experts! What is revealing is how seldom it is that expertise in general is attacked. Every group criticises experts on the other side but is happy with its own experts. Perhaps it's time to encourage people to think for themselves rather than always trusting someone else.

There's more to social change than beating the experts
You may be the sort of person who loves to see experts squirm, just for the sake of it. But most people have a different basic aim: they want to stop fluoridation or nuclear power, promote the biblical view of creation, or maintain military preparedness in the face of nuclear-winter-backed demands for disarmament. Stripping the experts bare is a means to a wider end.

Some critics believe that if they could just show the holes in the standard arguments, the orthodox policy would collapse like a house of cards. Some antifluoridationists, for example, think that if scientists would just look at some of the evidence for harm from fluoridation, they would reject it outright.

If only life were so simple! Even if you can demolish the orthodox view at the level of ideas, that may not stop the policy associated with it. To put it bluntly, ideas are seldom the foundation of policies. Instead, policies are usually decided on the basis of vested interests, or in other words power politics in the widest sense. Ideas are brought in to justify the policy. The ideas are part of the politics, not the foundation for a rational and disinterested decision.

For example, the standard theory for dealing with modern-day market economies is neoclassical economics. Many critics have pointed out flaws in the assumptions underlying neoclassical economics, and in addition critics have demonstrated mathematical flaws that throw the whole theory into question. Has this dislodged neoclassical economics? Hardly. The critics have been ignored and research and policy making go on pretty much as before.

At best, stripping the experts is part of a wider strategy for changing policies or practices. This wider strategy may involve mobilising public support, working through organisations, building a committed core of activists, using money or charisma to attract followers, or taking direct action in the form of occupations and strikes.

Tobacco companies use their economic strength to advertise, hire researchers, win over politicians through contributions and lobbying, and sponsor sports and the arts. Attacking the antismoking scientific experts is only a small part of their overall struggle. It helps, but it is not the key to defence of tobacco interests.

Antifluoridationists use their ability to mobilise a core of committed workers in order to produce leaflets and newsletters, send letters to newspapers, lobby politicians and organise public meetings. Attacking the profluoridation experts is part of this campaigning, but only one part. Local antifluoridationists sometimes have succeeded even when opposed by the united weight of orthodox experts. And sometimes they have failed even when their attacks on orthodox experts have been quite biting.

The reason for this is quite simple. Most people, including politicians and other powerful decision makers, pay only limited attention to technical experts. The health hazards of smoking were known for a long time before much action was taken against it. Other things, including the popularity of smoking and the economic interests behind it, inhibited any action. On the other hand, some of the early and effective opposition to fluoridation was based on minimal evidence of harm. Other factors, including antagonism to government-sponsored measures and to tampering with the water supply, were enough to trigger strong opposition.

So it is important not to overrate the importance of experts and expertise. Seldom are they as influential as they like to think.

Go ahead and attack the experts. It's a valuable experience and you'll learn a lot. But remember, it's only part of a wider struggle. Good luck.

What you're up against: endorsements

The experts can be persuasive. They can quote stacks of facts and figures. They can give all sorts of logical reasons. They can present plausible explanations. They can also pose difficult questions and point to awkward contradictions in any alternative view.

However, the biggest thing going for the experts is that they are thought to be the experts. People think the experts are right *because* they are the experts.

Of course, not just anyone is treated like an expert. You need to have degrees, links with eminent institutions, and ties with prestigious professional bodies. A Nobel prize helps!

The establishment has one great advantage: endorsements. Endorsements by prestigious experts. Endorsements by eminent professional bodies. The experts don't even need to offer evidence and arguments. They can just refer to endorsements.

If you oppose fluoridation, you're up against endorsement upon endorsement. Frank J. McClure in his book *Water Fluoridation: The Search and the Victory* uses endorsements. Here is a taste of his presentation.

"Fluoridation has been given official approval by virtually all national and international health and professional organizations:

American Dental Association (1962)

'The fluoridation of public water supplies is a safe, economical and effective measure to prevent dental caries. It has received the unqualified approval of every major health organization in the United States and of many in other countries.'"[1]

McClure then quotes endorsements from the American Medical Association, the American Association for the Advancement of Science, the American Federation of Labor and Congress of Industrial Organizations, the American Water Works Association, and the American Institute of Nutrition. He then lists by name 34 additional

---

[1] Frank J. McClure, *Water fluoridation: The search and the victory* (Bethesda, Maryland: U.S. Department of Health, Education, and Welfare; National Institutes of Health; National Institute of Dental Research, 1970), p. 249.

US and 15 British organisations that have endorsed fluoridation, followed by various additional statements, such as by the World Health Organization.

In the face of this apparent agreement by the experts, are you sure you're right and the experts are wrong? If you are against the experts, you will have relatively few, if any, endorsements compared to their prestigious backing.

Endorsements — and, more generally, the status and prestige of accepted authority — constitute the major difference between the establishment and the challengers. It is a big difference, but sometimes it is possible to overcome it.

Caution: experts are powerful and dangerous

Stripping the experts bare is possible, but not easy. Proceed with caution. Plenty of planning and preparation is essential. If the experts condescend to notice your criticisms, they may decide to crush you, more ruthlessly than you imagine.

To tackle the experts you need to study and practise. Often you will become an expert yourself. But you won't need to know everything the experts do. You become an expert in the weaknesses of the orthodox view and in ways to exploit those weaknesses. Sometimes that isn't so hard, and it can be fun too.

I can't guarantee that you will be successful. You might even decide, after studying the issues more carefully, that the experts are right after all. It happens sometimes.

Disclaimer

My aim is to *describe* how establishment experts can be attacked. I certainly don't personally *recommend* every one of these techniques. Indeed, I oppose dirty personal attacks and prefer calm, fair-minded discussions of issues. Unfortunately, there are lots of nasty attacks and all too few calm discussions. Therefore, it's important to *understand* the common techniques, even if you never use them, because you are likely to encounter them, whichever side you support.

1
Challenge the Facts

"Expert: one who can take something you already know and make it sound confusing." — anonymous

Experts usually rely on what they claim are "facts." These facts are tied together in "arguments," which are logically organised sets of statements that lead to a conclusion. For example, "Nuclear power is safer than other energy sources, as shown by the fact that not a single member of the public in a Western country has died from a nuclear reactor accident."

There are several ways to counter facts.

Challenge facts directly

Sometimes experts get their facts confused or just plain wrong, whether due to laziness, stupidity or exaggeration in the heat of debate. This happens more often than you might expect. Among experts, it is often not considered polite to make a big fuss over mistakes. Especially when a contentious public issue is at stake, experts band together. They are reluctant to publicly expose each other's mistakes, since it might hurt their cause.

In any case, when an expert does get the facts wrong, you can pounce.

Leslie Kemeny, a researcher in nuclear engineering, has been a leading proponent of nuclear power in Australia. In one case, he made a blatant mistake when he wrote "Despite the tragedy of 1946, international experts in radiation biology and genetics have not found an incidence of genetic malformation, cancer or leukemia amongst these people above that of the national average."

Kemeny's trivial mistake was to write 1946 instead of 1945, the year atomic bombs were dropped on Hiroshima and Nagasaki. Kemeny's big mistake was to say that cancer and leukemia did not increase. All nuclear experts say they have. (The evidence on genetic defects is not sufficient to show whether or not radiation has had an effect.)

Kemeny has written numerous articles promoting nuclear power, and even in the article with his blatant mistake, there were many other statements that would be harder to challenge. The key is to home in on the mistakes. This is what the critics of nuclear power did. They wrote letters to the journal *Engineers Australia* that had published Kemeny's article.

Once he was challenged, there were several responses that Kemeny could have made. One was to simply ignore the criticisms. This can work if the expert is prestigious and the critics aren't. All you can do in this case is keep rubbing in the expert's mistake.

Kemeny, however, decided to reply. On his trivial mistake about the year 1945, he limited the damage by admitting it. On trivial points this is an effective response. Kemeny could also have admitted his more serious mistake about cancer and leukemia, but this would have been a damaging admission. He could have tried to defend his statement, but in this case that would have been foolish, since he didn't have a leg to stand on. Another approach is to avoid and confuse the issue. Here is what Kemeny wrote:

"Within the limits of uncertainty associated with medical diagnostics nothing in this report ['The Delayed Effects of Radiation Exposure Among Atomic Bomb Survivors, Hiroshima and Nagasaki, 1945-1979'] and many others dealing with the issue ['the radiobiology of high level radiation'] which I have on file, changes the veracity of the third paragraph of my [article] which relates to the first and second generation progeny of the Hiroshima survivors."

If you can decipher this jargon, you will see that Kemeny has avoided mentioning his mistake, which was about cancer and leukemia, by defending what he had said that

CHALLENGE THE FACTS

was correct anyway. In other words, he has attempted to wriggle out of his mistake without admitting error.

There is a very important lesson here. When an expert makes a mistake, it should be hammered home relentlessly. In a further reply to Kemeny, one critic wrote that "Kemeny failed to contradict the charge" made against him, "but successfully obscured the issue by his rhetoric."

If the expert admits making an error, then the critics can harp on the fact that the experts do make errors. "What about the errors that haven't been exposed?" If the expert refuses to admit the error, then the critics can claim that a coverup is occurring.

In countering the facts, it is vital to use extreme care in picking the fact you are going to challenge. If the experts are used to debating, they will be careful about what they say, and not as vulnerable as you may think. But sometimes they trip up. When that happens, you should be ready to strike.

Point out counterexamples

Whenever an expert makes a generalisation, try to think of exceptions. These are known as "counterexamples". Pointing out counterexamples is one of the very best ways to argue against rules.

There are loads of counterexamples to those slogans on cigarette packets saying "Smoking reduces your fitness" or "Smoking kills." My neighbour was a chain smoker and lived to be 90, when she was hit by a bus. And then there are those people who never smoke, eat all the right foods and exercise fanatically, yet who look terrible and die young.

Creationism, the belief that the world and its life forms were created by God pretty much in the form we know them today, uses counterexamples to undermine the theory of evolution. For example, creationists point out that the fossil record contains no transitional forms, intermediate between different biological species. Nor are there fossil records of intermediate stages of new body features, such as bones or organs. For example, in the evolution from the fish to the amphibian, one might expect to find fossils showing development of the pelvis of the amphibians. There aren't any.

The best counterexamples are ones that people can readily recognise and that can't easily be explained away. Defenders of the orthodox position usually have to resort to complicated statistics or intricate theories in these cases, which is a real weakness.

"Explain" the facts

On many occasions, experts make statements that can't easily be shown to be wrong — but neither can they conclusively be shown to be correct. These facts are unprovable. On other occasions, facts are misleading because they are out of context or apply to only some situations. In these cases, a good response is to "explain" what is really going on and thereby expose the limitations of the expert's facts.

One of the standard statements made by supporters of nuclear power has been that not a single member of the public has ever died due to an accident at a nuclear power plant. (This was before the Chernobyl accident which directly killed civilians in an obvious way.) A good way to respond is to explain this fact by describing what it doesn't say.

• Many thousands of workers (who don't count as "members of the public") have died in accidents in nuclear power plants and other facilities in the nuclear fuel cycle;

• Millions of people have been exposed to low-level ionising radiation from nuclear facilities. Many of them may have died due to cancer as a result, but there is no sure way to know whether any particular cancer was due to one cause or another.

In other words, the pre-Chernobyl fact about deaths from nuclear accidents was unprovable. Simply by "explaining" what the fact doesn't say, you have exposed this, and also made some good points. The expert's fact then is revealed as misleading.

Point out uncertainties

Associated with any fact is a degree of uncertainty. The number of suicides in the United States in 1984 may be listed as 29,286, but a proper assessment of that figure should include all likely sources of error and interpretation, including wrong diagnoses, varying interpretations (for example, of vehicle accidents), cover-ups (by relatives) and social customs (is death by alcoholism self-inflicted?). Even hard scientific data, such as the number of counts on a scintillation counter, are uncertain due to possible machine errors, errors in transcribing data and mistakes such as measuring the wrong sample.

The science involved in many controversial issues is subject to a large degree of uncertainty. Scientists, naturally enough, often prefer to emphasise the central result and to downplay the sources of uncertainty. Often a single figure or conclusion is stated, with no error limits mentioned. This is a serious vulnerability that can be exploited.

Since 1982, there has been a flurry of scientific research on the global climatic effects of nuclear war, in particular the "nuclear winter" effect of sudden cooling. Most of these studies are based on computer models of the effects on the atmosphere of dust raised by nuclear explosions and from soot from fires started by the explosions. These studies are beset by a multitude of uncertainties. The targets for nuclear weapons are uncertain. Whether nuclear missiles will work and be accurate is uncertain. The amount of dust and soot generated is uncertain. The extent of coagulation of particles in the atmosphere is uncertain. The computer models do not incorporate every possible effect, and so their results are uncertain. One could go on and on raising uncertainties.

This is what some of the critics of nuclear winter theories have done. The scientific orthodoxy has been that a substantial global cooling is likely. By emphasising the uncertainties, the critics suggest that nuclear winter is uncertain itself.

A similar approach has been taken by the critics of the medical establishment view that smoking is linked to lung cancer and other diseases. The critics point out that there is no definite proof, at a microscopic level, that smoking has directly led to cancer. At best there is statistical evidence to support a theoretical model for the cause of the cancer. What about other models? Perhaps people who are prime cancer victims anyway are just the sort of people who enjoy smoking. Perhaps there are a range of environmental factors that independently cause certain people to get cancer and to smoke. In other words, there is an inherent uncertainty or lack of conclusiveness in the theory and statistics behind the usual claim that smoking causes cancer.

Defenders of orthodoxy always present their views as more established and secure than is actually the case. They emphasise the main results in public and restrict discussion of uncertainties to specialist articles. So it is almost always effective for critics to draw attention to the uncertainties. The more uncertain the result, the more it seems that the view of the critics may be just as likely as that of orthodoxy.

Point out other facts as a distraction

A tried-and-true method of challenging uncomfortable facts is to ignore them and to draw attention to other facts or topics that are more congenial. Valuable insight into this technique can be gained by watching how good politicians respond to awkward questions.

But pointing out other facts is a more serious method than simply avoiding the issue. Experts try to define the issue in terms that make their own expertise central. If critics accept this, they are trapped into confronting the experts on their home territory. It makes much more sense for the critics to raise their best possible arguments, even if this means changing the topic entirely.

In the nuclear power debate, the nuclear experts have concentrated on issues of reactor safety and disposal of long-lived radioactive waste; these are the areas where nuclear expertise can be brought to bear most powerfully. Opponents have raised all sorts of other issues: terrorism, proliferation of nuclear weapons, uranium mining on indigenous land, threats to civil liberties, energy efficiency and renewable energy sources, etc. These "other issues" include, more obviously, social and political dimensions, and thus are harder for nuclear scientists and engineers to confront.

In the debate over adding fluorides to public water supplies, proponents emphasise the large reduction in tooth decay and the lack of any demonstrated hazards. This is the orthodox position. Rejecting this, opponents have claimed that all sorts of physical ailments are due to fluoridation, including kidney damage, digestive problems, allergies and intolerance reactions, Down's syndrome, skeletal fluorosis and cancer, among others. Naturally, the opponents demand conclusive proof that fluoridation is not causing any of these problems. As soon as any proponent has described any research showing fluoridation is probably not responsible, the opponents have raised half a dozen other possible diseases.

The tobacco industry and other defenders of smoking have put up a valiant struggle against critics, especially in criticising the medical claims about death and ill health. But in many cases they have found it is better to move to a different defence: civil liberties. Selling cigarettes is legal, and therefore there should be no special measures taken against smokers. Shifting to the human rights argument avoids the more difficult area of smoking and health, and also avoids direct confrontation with the medical experts, who can claim no special expertise on issues of civil liberties.

Raising "other facts" or shifting the focus of debate is of central importance in challenging experts. To reiterate, the experts, as might be expected, concentrate on the areas of their expertise, usually the most technical areas that others can't understand. Confronting the experts on their home ground is not to be undertaken lightly, and is a job for only the most experienced and confident "counterexperts" (the experts on the other side).

The aim should be to shift the debate from where the experts are strongest to where they are more vulnerable. If the pronuclear experts are nuclear scientists, then shift the debate to social and political topics such as terrorism. If the profluoridation experts are dentists and doctors, then shift the debate to the issue of individual rights that are violated by compulsory medication. If the antismoking experts are doctors, then shift the debate to the right to smoke.

Summary

There are a variety of ways to counter facts presented by the experts. Most direct is to confront them as wrong, inconsistent, impossible or unprovable. The opportunities for such direct attack are usually limited. So when the occasion does arise, ram the point home relentlessly.

More effective are counterexamples to statements by experts. It is worth spending a lot of time investigating possible counterexamples, and then choosing the best ones, especially those that are direct and emotionally appealing. One good counterexample often can undermine an entire body of statistics.

"Explaining" facts is effective if there is an opportunity. By exposing what wasn't said by the experts, sometimes you can turn the hostile facts into assets.

Complex theories often can be attacked by pointing to uncertainties. Referring to uncertainties always helps if you are on the weaker side in terms of scientific backing.

Finally, raising other facts and changing the focus of debate is a crucially important tactic. The experts always try to peg the issue on their strongest points and their areas of expertise. You should try to pull the debate towards *your* strongest points and the areas where the experts have least formal authority.

Preparation and training to counter the facts

In order to be effective in countering the facts, you need to study the issue carefully and be absolutely sure of what you are doing. Usually there are some counterexperts, people who will have studied the issue in depth but disagree with the orthodox stance. Sometimes these are former orthodox experts who have changed their minds. Reading their writings and hearing them debate the issue are excellent ways to learn how to undermine the experts.

It is also vital to study the arguments of the experts themselves. Don't rely on what the critics say that the experts say. You need to read and listen carefully to what the "other side" says, seeing exactly what they say and how they say it. If you've also studied the critics, you should be able to see weak points, think up counterexamples, and notice what issues are being avoided.

Experts may make mistakes because they are arrogant, overconfident, careless or rushed for time and don't bother to check their details or spell out qualifications and uncertainties. Or else they may only check with others just like themselves.

Whenever you are countering the facts, it is essential that you check and doublecheck your facts before going public with them. A sympathetic critic, or an expert not involved in the debate, often will be willing to help you. If you neglect to check and doublecheck, you are likely to regret it.

It is also important to use a sensible style. This doesn't mean the anaemic and arid scientific jargon of most experts. But it is wise to avoid styles that look amateurish. Avoid CAPITAL LETTERS and lots of exclamation points!!! Avoid super-dramatic claims and announcements and wild allegations of fraud and lying. Even if everything you say is true, it is usually more effective to avoid excesses of rhetoric.

"In fact, the intentional avoidance of all colorful or emotional words is itself a powerful dramatic choice — one of the oldest known. It is designed both to inspire automatic trust and to lend additional, unearned weight to every word uttered. As a device, it can usually be used effectively only by those whose previous reputation, rank, office, or position projects an aura of its own before they arrive."[1]

---

1 Henry M. Boettinger, *Moving mountains or the art and craft of letting others see things your way* (London: Macmillan, 1969), p. 34.

2
Challenge Assumptions

"An expert is one who knows more and more about less and less." — Nicholas Murray Butler

Every scientific theory and every set of beliefs is built on assumptions. One powerful way to challenge the experts is to attack their assumptions.

This is often more effective than challenging their facts. The facts are things that they develop, understand and deploy in arguments. So experts are usually at their strongest in debating the facts (at least the ones they bring up).

By contrast, experts have fewer choices in the assumptions that underlie their arguments. The assumptions are the soft underbelly of many an expert's apparently solid body of evidence.

The assumption might be that mathematical calculations can be used to compare people's values, or that governments always have the best interests of the people at heart, or that experts are never corrupt.

Sometimes all you need to do is to expose the assumption for all to see. It may be wrong, ludicrous or distasteful to most people. Other times you may need to show why the assumption is a bad basis for trusting the expert.

It is especially important to concentrate on assumptions that are never revealed by experts. If they feel free to state an assumption openly, it is probably not very dangerous for them.

Actually, there is not a rigid distinction between facts and assumptions. Many assumptions can be backed by facts to some degree. The key thing is that the arguments of the experts depend vitally on assumptions, whereas a few facts here or there don't make that much difference.

Expose wrong assumptions

Assumptions underlie everything we do and say. What those assumptions *are* is sometimes obvious but more commonly obscure or contentious. To say that an expert's assumption is wrong is to say that it is illogical, easily refuted or just plain ridiculous. But it is remarkably difficult to expose wrong assumptions. They may be hidden or, worse, connected to popular prejudices. The challenge is to bring assumptions out into the open in a way that demonstrates their absurdity or foolishness.

Proponents have often used the spread of nuclear power as an argument *for* nuclear power. But there is a missing link, a hidden assumption. The assumption is that nuclear power must be a good thing, because so many countries are adopting it.

More generally, the unstated assumption is that adoption of a technology means the technology is beneficial. The easiest way to counter this assumption is to offer counterexamples. Lots of other technologies have spread around the world that are undesirable, such as weapons of mass destruction or aerosol sprays that damage the ozone layer.

A more general form of this same assumption comes in the form of the statement "Scientific progress can't be stopped" or "You can't stop progress." Claims of this sort have been made about nuclear power, fluoridation, computers, genetic engineering and a host of other technologies. The assumptions involved are that the technology in question represents progress, and that it is impossible to stop its rational introduction. (So why bother, you fools?)

Obviously, if you are opposed to the technology, you don't think it is progress. The trick here is to show that you are promoting *real* progress.

Once again, pointing out counterexamples is an effective way to respond. There are plenty of technologies that have not been widely adopted, such as the supersonic transport aircraft and margarine in squeeze bottles. Technology *can* be stopped. More importantly, there are all sorts of "bad" technologies that have spread around the world, such as chemical weapons and implements of torture.

Another response is to ask, "If progress can't be stopped, why is it necessary to push so strongly for nuclear power (or whatever)?"

The claim "You can't stop progress" is worth studying carefully because it is encountered so often. It begs the question of what is progress, and also hides the fact that introduction of new technologies requires ongoing efforts by numerous people. Technologies don't just suddenly walk in the door by themselves and join the household; they have to be planned, developed, manufactured, sold and used. The claim about not being able to stop progress is especially annoying because it is blatantly biased and ridiculous and yet so persistently raised.

Expose flawed assumptions

In many cases, assumptions are not obviously wrong or foolish, but still can be criticised as weak or insufficient for the task required of them.

The theory of evolution assumes that the process of genetic mutation is the basis for explaining the development of new species. Yet mutations are random, and it is accepted that the vast majority of them are harmful to survival. The critics of evolution say that since the process of mutation is blind to what happens, it is not able to explain the dramatic developments in new life forms. How can accidental genetic change lead to new life forms? Analogies are useful here. How likely is it that random changes in the letters of Shakespeare's *Hamlet* will result in the memoirs of Richard Nixon?

Present counterassumptions

A counterassumption is a completely different assumption that can be used to arrive at completely different conclusions. Presenting a counterassumption is a good way to expose the limitations of an assumption. (A counterexample, by contrast, can be used to attack an assumption but doesn't present an alternative.)

Creationism contains within its name a counterassumption to biological evolution, namely that direct creation is a way to explain the present diversity of life forms on earth. The process of genetic mutation plus natural selection can explain many features of life only with difficulty, requiring mathematical calculations, assumptions about sudden bursts of evolutionary activity, and assumptions about transitional forms that do not appear in the fossil record.

The assumption about creation is, by comparison, straightforward. Complex life forms are the way they are because that is basically the way they were created. This counterassumption is effective in highlighting the many assumptions underlying conventional evolutionary theory.

The medical theory of overuse injuries is that when muscles or joints are used too often or with excessive strain, organic injury results. This can happen at work, as in the case of tenosynovitis from typing, or from recreational activities, as in the case of tennis elbow. The assumption here is that pain and disability are due to physical damage to parts of the body.

The critics of the medical theory use an alternative assumption: that nonphysical factors are behind most reports of overuse injury. They rely on theories of occupational neurosis or conversion hysteria, which in essence say that workers develop symptoms as an unconscious way to get out of work or to escape personal psychological conflicts. These psychological mechanisms may combine with pain due to normal muscle fatigue and result in the perception (and reality) of unrelenting pain and disability.

The counterassumption behind this critique of the medical view of overuse injury is that psychological factors are crucial in the explanation.

Counterassumptions are powerful because they do two things at once: expose actual and potential limitations in the assumptions underlying the standard view, and provide the basis for an alternative view. This is also their weakness. Counterassumptions may themselves be challenged and exposed as flawed. Simply having a counterassumption is not enough. It should be one that can stand up to intense scrutiny.

But in challenging orthodoxy, the alternative doesn't have to be perfect. The object should be to throw the standard view into question and open up the issue for wider debate.

Deny the implications of the assumption

One way to challenge an assumption is to say "so what?" If you can say this and back it up, it means that there is a hidden assumption that you are implicitly challenging.

The publicists of nuclear winter have made a big fuss about the likely catastrophic effects of nuclear war: as well as massive death from blast, heat and fallout from nuclear explosions, there is likely to be freezing and mass starvation in most countries not hit by weapons. This may seem hard to counter on the basis of science. But when it comes to policy implications, a good answer is, "so what?" "We knew that nuclear war was terrible. The main aim is to prevent nuclear war, and the nuclear deterrent is the best way to do this."

In fact, most people believed that nuclear war would kill nearly everyone even before the nuclear winter theory was developed. Nuclear winter doesn't change the imperative to avoid nuclear war.

The reason that this response works is that the scientific theory of nuclear winter has had attached to it a political assumption: the larger the effects of nuclear war, the stronger the argument for nuclear disarmament. It is this political assumption that is challenged by the response "so what?" It says, in effect, that bigger effects from nuclear war do not necessarily lead to the conclusion that nuclear disarmament is warranted.

Is there a genetic basis to variations in human intelligence, or is intelligence mainly shaped by environmental factors? Those arguing that environmental factors are more important usually are the ones arguing for measures to promote equality in society, such as equal access to top jobs for women and ethnic minorities. They oppose theories based on genetic inequality as providing a scientific basis for social inequality.

But there is another response to the scientific argument that human intelligence is mainly genetic in origin: "so what?" Why not promote social equality anyway? People who are blind are not usually left to fend for themselves; special provision is made for them to live as normal a life as possible. People with poor eyesight, who have to wear glasses, are hardly penalised at all. Technology allows many handicaps to be overcome, including ones that in previous eras would have meant automatic poverty or death. People no longer have to be fit enough to capture wild game or sow their own crops. In the modern technological society, it can be argued, genetic inequality is no excuse for social inequality. If people decide that equality is a good thing, then technology can be designed to make it possible.

Again, this "so what?" response exposes a political assumption attached to the scientific evidence, namely that genetic inequality justifies social inequality. Exposing this political assumption then allows it to be challenged.

Summary

A good way to undermine the arguments of experts is to attack their underlying assumptions. The key step is to expose the assumptions, which are usually hidden behind the facts and the conclusions. Often, just by exposing certain assumptions, they are revealed as wrong, flawed or objectionable.

Another approach is to develop and promote counterassumptions, namely assumptions that are different from the standard ones. This throws the standard assumptions into sharp relief and also presents an alternative way forward. It has the possible disadvantage that the counterassumptions may come under attack.

Finally, it is possible simply to deny the implications of assumptions by saying "so what?" This helps to expose the political side to what is ostensibly a scientific or specialist issue.

How to uncover assumptions

Uncovering the assumptions underlying the views of experts is a challenging operation but well worth the effort. There are several ways to go about this.

One straightforward approach is to ask experts themselves. Their understanding of their own assumptions is a good place to start, even if these are usually put in the best possible light. Every expert will claim to have the best interests of society (or of science) at heart, of course.

It is essential to consult counterexperts. They are the ones who have made a detailed study of the expert arguments and their weaknesses. Part of the weaknesses will be in the assumptions.

But one problem with experts and counterexperts is that they tend to be caught up in the game of expertise and the rules by which it is played. So it is hard for them to see the really basic assumptions, especially the ones of central significance.

One way to tackle the issue yourself is to set down the facts that the experts claim are the basis for their view, and say, "Do I come to the same conclusion as they do?" If so, you may disagree only about facts. But if not, assumptions are almost certainly involved.

For example, critics of the nuclear winter theory may accept that computer models show a massive cooling in the aftermath of nuclear war. Why don't they accept this result? The issue goes back to the limitations of computer models. Every researcher accepts that computer models do not incorporate every effect. The divergence appears in assessing the implications of this. The nuclear winter supporters think that more complete models will vindicate their original results, showing effects nearly as bad, if not much worse. The critics think that including certain extra effects in the models may well result in a much less significant effect. So the key assumption concerns whether the original models give representative results, or whether they give exaggerated results.

Another approach is to think up all sorts of counterassumptions and see where they lead. If they lead to the same sorts of conclusions as the experts, then the assumptions probably aren't crucial. But if you get different conclusions, then you may have hit on a point of attack. Of course, you need to be sure that your counterassumption is one you are able and willing to defend.

A stimulating way to test for assumptions is to imagine different types of societies: extreme equality or inequality; highly centralised or extremely decentralised; monolithic in beliefs or very splintered; ruthlessly competitive or extremely generous to the disadvantaged; militaristic or peaceful; ethnically diverse or completely uniform; comsumer-goods oriented or personal-values oriented; authoritarian or tolerant; rich or poor. Then consider the issue in question in the context of several hypothetical contrasting societies. For example, in which sorts of societies would fluoridation be considered suitable? In which sorts of societies would occupational overuse injuries be more likely to arise? You may find that the arguments of the experts depend on society having certain characteristics. If you can argue that this is not the sort of society that is desirable, then you can confront the expert's assumption.

Appendix: Assumptions made in mathematical models

These days, more and more experts are backing their claims by referring to mathematical models. Back in the early 1970s, the limits to growth models received widespread attention. Using a simple computer model of factors including population, resource use and pollution, the modellers claimed that economic growth would inevitably lead to disaster. This was the same as what a lot of environmentalists were saying anyway, but the limits-to-growth people were backed up by that mysterious thing called a mathematical model.

A large fraction of theoretical work done in the sciences uses mathematical models. This means representing things in nature or society using symbols that are manipulated mathematically. The mathematical result is then interpreted in terms of what it means for real life.

An example is r = pc.

Here, r stands for the risk associated with a technology, p represents the probability of a particular hazard, and c is the size of the consequences. For example, if the chance of a Chernobyl-style nuclear reactor accident is one in a million per year per reactor and causes 20,000 deaths (immediate and longterm), then for the United States r = 0.000001 accidents/reactor/year × 100 reactors × 20,000 deaths/accident = 2 deaths per year. By comparison, the same calculation for cars in the United States might look like r = 0.05 accidents/car/year × 100,000,000 cars × 0.01 deaths/accident = 50,000 deaths per year.

In any calculation using a mathematical formula or model, it is always possible to criticise the actual numbers inserted into it, such as the figure for the number of accidents per reactor per year, especially in a case like this where there isn't enough evidence to provide a firm figure. But often it is more effective to attack the modelling process itself by exposing the assumptions underlying the model.

First, there are assumptions involved in representing reality by symbols. The symbol c in the formula r=pc is supposed to represent consequences. Deaths are relatively straightforward to calculate, but how can one evaluate consequences such as disability, pain and fear? A close look at the symbols in any mathematical model will reveal all sorts of inadequacies in arbitrarily measuring quantities, grouping different things together, and excluding qualitative effects.

Second, there are assumptions involved in how the symbols relate to each other. The formula r=pc assumes that it doesn't matter whether the risk is due to many small events or a few enormous ones. Yet many people treat these quite differently. For example, suppose the chance of a nuclear war that kills everyone on earth is one in a million per year. The r = 0.000001 human exterminations/year × 5,000,000,000 deaths/extermination = 5000 deaths per year. This is one tenth the "risk" due to traffic deaths in the United States alone, yet many people would consider human extermination to be much more serious! The risk formula r=pc doesn't allow for this.

(This flaw is related to a fundamental flaw in utility theory, which is widely used in economics. Comparisons of welfare are done on the basis of "utilities," which are measures of benefit or loss that are compared to each other using probability-based tradeoffs. The trouble is, a single infinite utility value wrecks the entire model. And for many people, their own death is non-negotiable. This creates an infinite utility and the model breaks down for everyone. This is an example of how a simple counterexample can undermine a seemingly well-established model.)

Third, models always leave some things out. If they didn't they wouldn't be models — they would be reality itself. The strength of models lies in their ability to represent key features of reality while leaving out messy but unimportant details. Because of its simplicity, a mathematical model can be manipulated much more easily than reality. But this strength of models is also their weakness.

One thing commonly left out of models is other ways of doing things. The formula r=pc can be used to compare nuclear reactor accidents with automobile accidents, but does not lend itself to looking at the effect of energy efficiency (reducing the need for nuclear or any other energy) or the spread of telecommuting (people using telecommunications to work at home and reduce automobile use). Once the risk formula is brought to bear, the implication is that the sources of risk are more or less inevitable.

Perhaps the most crucial assumption in any model is that it incorporates the most important features of the reality being studied. You can expose the model as deficient or even immoral if you can come up with things that are not included in it or, even better, with things that cannot possibly be included in it.

Fourth, modellers make assumptions when they interpret the results. The care taken in designing and using the model is often forgotten in making grand pronouncements based on the results. In many cases an extra political assumption is involved, as in the case of the nuclear winter models described earlier.

In the case of r=pc, the cautious interpretation is to compare only those risks arising from similar causes. This might be the death rate from auto accidents in two different cities. Even here it is difficult to draw conclusions, since differences could be due to the condition of roads, weather, the servicing of the cars, policing or drug-taking.

Broader interpretations are even more open to challenge. It is common for defenders of nuclear power, for example, to use a sophisticated version of r=pc to compare risks from different energy technologies, or to compare nuclear reactor accident risks with risks from driving cars or being struck by lightning. Their standard argument is that because people accept risks from cars and because the risk from nuclear power is much less than that from cars, therefore people should learn to accept the risks from nuclear power.

This sort of argument relies on extra assumptions added in the interpretation stage of the risk calculation. The comparison assumes, among other things, that the risks are similar in nature and origin, that the benefits associated with each risk are similar in magnitude, and that there are no alternatives that have similar (or larger) benefits and lower risks. You can challenge each of these assumptions, tossing in a few counterexamples to dramatise the points.

In this and many other cases, models are used as a mask for political viewpoints. Mathematics is supposed to be rigorous, and so most people expect mathematical models to be much more objective than a set of opinions strung together. Undoubtedly, mathematical models do have their effective uses in all sorts of fields. When they become most subject to attack is when they are used in areas that have social implications. In these areas, modellers may unconsciously build in assumptions that give results they, or whoever funds them, find useful.

Many mathematical models are intimidating in their complexity. Often they include hundreds of variables and equations, and incredibly difficult and lengthy calculations, usually carried out by computer. Complex models are harder to challenge because they are hard to figure out and because they impress many people as being the product of objective science.

Even the most complex model is subject is inaccuracy due to uncertainties in the data and mistakes in calculation. If you can find important shortcomings in these areas, well and good. But a more fundamental challenge is to look at the assumptions in the model itself: in the key symbols, in relating the symbols to each other, in interpreting the results, and in things left out of the model.

The biases and shortcoming of earlier models can be used as evidence against trusting new ones. The initial limits-to-growth models showed an inevitable collapse of civilisation if economic growth continued without interruption. Critics later did calculations that showed that the model was operating far outside the proper range of many of its variables (symbols); with some fairly minor changes, the critics showed that the model didn't necessarily result in collapse at all. Others later made assumptions about sharing wealth that also led to different results.

Pro-nuclear energy modellers at the International Institute for Applied Systems Analysis in Vienna came up with results showing the necessity for nuclear power. Critics investigated the model in detail and claimed that for all its massive complexity, the core was really very simple, and also flawed: it generated results similar to arbitrary input data, and this input data had been skewed to give pronuclear results.

3
Discredit Experts

"No lesson seems to be so deeply inculcated by the experience of life as that you never should trust experts." — Lord Salisbury

The credibility of experts as experts depends to a surprising degree on their personal credibility as individuals. If they are seen as honest, concerned members of the community, their views will carry more weight that if they are revealed as greedy and arrogant snobs who beat their children. Yet, logically speaking, the quality of an expert's views should be treated as independent of whether the person is nice or nasty.

If you can discredit experts personally, by whatever means, you will probably also destroy the credibility of their expertise. Logic aside, an expert exposed as a thief will not be as effective as an expert noted for philanthropy, even though their expertise may be on something entirely separate such as astronomy or fluoridation.

It is a risky business to go about exposing the darker side of the private lives of experts. Personal attacks sometimes backfire, since many people recoil against attacks that are seen as unfair or intrusive. Therefore, the most effective attacks on experts as people are those that target their personal limitations as experts: mistakes, inconsistencies and vested interests.

Expose failures

If you can show that an expert has made a mistake at any time on any issue, you can use this to argue that experts shouldn't be trusted now, on the current issue. Exposing failures is a powerful way to discredit experts. Nothing they can do in response is really effective.

If they admit the failure, you can ask why the current case is any different. If they refuse to admit the failure, but argue that their view was correct at the time (or is still correct), you can contest the point vigorously. This swings the argument to the credibility of the expert on a point that you, not the expert, have selected. (So be careful in your selection.) Finally, the expert can decline to respond at all, in which case you should keep hammering the point. Silence implies guilt, doesn't it?

Several of the scientists leading the promotion of nuclear winter have a history of making warnings of environmental disaster. Most prominent among these is Paul Ehrlich, professor of population biology at Stanford University, who, along with Carl Sagan, has been the leading populariser of nuclear winter. Ehrlich has a history of predicting environmental doomsday. His best-selling book *The Population Bomb* warned of impending disaster from overpopulation. Although population has continued to increase, there has not been a major social or environmental collapse. Arguably, Ehrlich was wrong, or at the very least overdramatised the dangers, in his earlier warnings. Why should nuclear winter be any different? Why should he be believed this time when he cries wolf?

(Of course, Ehrlich would argue that his population predictions were *not* disproved. Even if they were, that's no automatic reason why his nuclear winter views are less credible. But exposing his (alleged) false prediction nevertheless is a potent form of attack.)

The claims of supporters of nuclear power have gone astray many times. One of these claims is that nuclear power will not lead to proliferation of nuclear weapons, because of the safeguards of the Non-Proliferation Treaty and the inspections by the International Atomic Energy Agency. But India exploded a "nuclear device" in 1974. What happened to the safeguards then? In 1981 Israeli jets bombed a nuclear complex under construction in Iraq. Iraq had signed the Non-Proliferation Treaty; didn't the Israeli government trust the safeguards? There is also strong evidence that the Israeli government has had

DISCREDIT EXPERTS 35

nuclear weapons for years, and obtained uranium from the United States in 1968 by an illegal transfer. What about the safeguards? And where were the safeguards when the Pakistan government in the 1980s bought parts from European suppliers for a uranium enrichment plant?

These sorts of examples have the greatest impacts when juxtaposed with statements from nuclear experts about the confidence we should have in safeguards. A good collection of quotes and counterexamples is invaluable in a debate.

(Pro-nuclear experts can reply in each case that safeguards were *not* breached, or that breaches do not set a precedent relevant today. But by attacking their (alleged) failures, you put them on the defensive and undermine their credibility.)

There is also the good story about a satellite being launched by the US government. It was powered by the highly toxic plutonium-238. After concern was expressed about the potential danger, an official was quoted as saying the risk of failure was one in a million. After launch, the rocket disintegrated and sent the plutonium-238 into the atmosphere. You can quote this example whenever someone suggests trusting the experts.

The supporters of Immanuel Velikovsky's theories can point to many failures by the orthodox scientific community. In 1950, Velikovksy's book *Worlds in Collision* was published. Velikovsky claimed that Venus was a recent planet, having erupted from Jupiter and passed near the earth only a few thousand years ago. Velikovsky's unorthodox view suggested that Venus should be hot, whereas conventional science at the time expected that the surface of the planet would be cold.

Later, various interplanetary probes found that the surface of Venus was very hot indeed — an enormous surprise. Scientists tried desperately to explain this, invoking a "supergreenhouse effect." Even if the temperature of Venus can now be explained by conventional science, it is still true that the experts were badly wrong. Furthermore, it is relatively easy to come up with an explanatory theory *afterwards*. Scientists know that whatever the evidence shows, a theory can be generated to explain it. It is much harder to come up with a successful prediction. Velikovsky's theory provided the correct prediction at the time — not orthodox science.

(Critics of Velikovsky say that he was lucky, that he got the right answer for the wrong reason, and that he made so many predictions that some were bound to be right. None of this gets around the fact that the experts were badly wrong.)

Expose inconsistencies

If you delve into the past, you can often find that experts change their expert views, sometimes in an embarrassing way. The more prominent the expert, the more likely their views will be on record to be used against them.

Sir Ernest Titterton and Sir Philip Baxter were the leading scientist proponents of nuclear power in Australia from the 1950s through the 1970s. In the 1970s they were extremely active in writing and speaking in favour of nuclear power and uranium mining. To counter the argument that nuclear power was leading to proliferation of nuclear weapons, they presented the usual pronuclear view that the Non-Proliferation Treaty and the various safeguards, plus technical difficulties, made proliferation virtually impossible.

This wasn't always their view. Back in the late 1960s, before Australia had signed the Non-proliferation Treaty, Titterton and Baxter each criticised the treaty, basically because they wanted to keep open the possibility for Australian nuclear weapons. Baxter wrote that nuclear power in Australia would provide a basis for quickly obtaining nuclear weapons if required. Titterton said that the Non-Proliferation Treaty was "a worthless and ineffective bit of paper." In short, they tailored their views to the tenor of the times: when there was a possibility for Australian nuclear weapons, they acknowledged the link with nuclear power; when Australian nuclear weapons were ruled out and proliferation became an argument against nuclear power, they touted the power of safeguards.

To expose an inconsistency such as this is to attack the credibility of the expert on any issue, because it suggests that any view they express may be simply an argument of convenience. This example about nuclear power and nuclear weapons is also good because it highlights a current and topical issue, not just an obscure inconsistency of historical interest only.

Most people ignore their own inconsistencies and are not even aware of most of them. Experts are no different. To find inconsistencies you will probably have to dig into past records and perhaps do other detective work. The more prominent the person, the more likely it is that you will find something.

Attack the relevance of credentials

Credentials are perhaps the most powerful advantage held by experts. People who have PhDs, people who are professors or research scientists, people who have written scholarly articles or books, people who are official advisors to governments, have a great advantage. Many people, without any further evidence, will assume they must know what they are talking about. Your aim is to show that they don't, or don't necessarily.

Credentials are symbols of knowledge and competence. Your aim is to show that the symbols are empty.

One of the most effective ways to attack credentials is to show that they aren't relevant to the issues at hand. This is very often the case and, if you think about it, shouldn't be surprising. Most experts are technical experts: they have credentials in areas such as physics, biology, medicine or law. On issues that have wide social relevance, these credentials are not so relevant. On a decision about energy policy or health policy, what relevance is it to know quantum electrodynamics, to be able to manipulate genetic materials, to perform open heart surgery or to understand patent law? Not much.

Your aim is to expose the lack of relevance of the credentials of those on the other side.

The scientists who developed and promoted the theory of nuclear winter are mostly atmospheric scientists like Carl Sagan and ecologists like Paul Ehrlich. Their formal credentials cover the computer models used to calculate changes in temperature and weather patterns due to a nuclear war, and assessments of the ecological effects of these changes. Although nuclear winter models can be attacked even in these areas, it is hard to attack the credentials of the nuclear winter scientists.

But nuclear winter calculations have involved much more than atmospheric science and ecology. The models start off with assumptions about how many nuclear weapons are exploded and where. This is home ground for strategic theorists; atmospheric scientists and ecologists have no special expertise here.

The modellers typically claim that more people will die of starvation than from the direct effects of nuclear weapons. But the likelihood of starvation depends on social and political responses to nuclear war, such as the degree of panic, transport of food supplies to population centres, maintenance of energy services and organisation of migration. These are not areas where atmospheric scientists or ecologists have any special expertise due simply to their training.

Finally, several of the nuclear winter theorists, in particular Carl Sagan, have drawn political conclusions. Sagan argues that "deep cuts" in nuclear weapons stockpiles (90% or more) are essential to avoid the possibility of nuclear winter.

But Sagan's credentials provide no special basis for pronouncing on the implications of nuclear winter effects. The critics of nuclear winter, including the US Defense Department, argue that their policies are the most effective in preventing nuclear war and hence the possibility of nuclear winter.

This example shows that even if all the experts in the field are united in their views, they can still be attacked by showing that their field is not fully relevant to the issue at hand.

The credentialed proponents of fluoridation are largely from the dental profession and, to a lesser degree, from the medical profession. The dentists are easier to attack. The antifluoridationists simply point out that dentists may be experts on how to fill cavities, but their training gives them no special knowledge about the non-dental effects of fluoride, such as skeletal fluorosis or cancer.

The credentials of doctors would seem more relevant, but they can be attacked as well. To determine whether or not a small percentage of people are suffering from a fluoridation-induced health problem requires special statistical skills. The specialist required is called an epidemiologist. Most doctors have no special training in this area.

If these sorts of attacks on credentials are not enough, the antifluoridationists, as a last resort, point out that dental and medical expertise is irrelevant to the health and public policy issues of introducing fluoridation. The issue of individual rights comes up here: should people have a right to water without added fluoride? Dentists and doctors have no special training in the study of individual rights versus community benefits. So why should their views be given any special consideration?

Sometimes other experts come forth to fill the gaps left by the narrow specialists. These days there are experts in public policy, in bioethics, in analysing scientific controversies, and anything else you can think of. Nuclear experts can be attacked when they go from assessing the *likelihood* of the risks of nuclear accidents to statements about the *acceptability* of the risks. But an expert on risk assessment can be brought in to pronounce on how decision making concerning risks should occur.

The proliferation of experts into all sorts of areas poses a problem for critics of experts, but so far the problem hasn't become too serious. To start with, the new-fangled experts aren't taken nearly as seriously as the traditional experts in well-established fields. One reason for this is that experts in fields such as ethics and policy-making deal in areas where lots of people feel that common sense is good enough. These fields haven't yet been taken over by experts. Nor are they likely to be in the near future.

In addition, the new-area experts are less likely to be totally aligned with the established experts. It is in their interests to be responsive to a range of interest groups. Statisticians are unlikely to back fluoridation as readily as dentists. It is to the advantage of statisticians, as a group, to be independent of any single outside group — to increase their own status, of course.

When you question the relevance of credentials, try to get the experts to defend themselves. The more the focus is on possible limitations of the experts, the better.

Expose vested interests

If the experts have a financial interest in what they promote, exposing it can be very damaging. If they have an ideological axe to grind, exposing it can be damaging. If they have status to protect, exposing this can be damaging. In fact, exposing any sort of an interest on the part of experts is an extremely effective way to attack them.

In principle, experts can be unbiased in their opinions even if they are receiving money, promotions and invitations to high society as a result of expressing them. But most people don't believe it. If you expose any sort of vested interest, most people will assume there is bias. They might be right too.

Financial interests are probably the most damaging type that can be revealed, at least in Western capitalist societies. In 1975, Charles Schwartz documented the corporate connections of a large number of leading pronuclear scientists and engineers. Out of 32 signatories to a pronuclear statement, two worked for corporations, four had been consultants to major corporations, 11 had links with the Atomic Energy Commission, and 14 were members of the boards of major corporations. The message was clear: these scientists were not truly independent experts. Their views were compatible with their pocketbooks.

Antifluoridationists have tried to pin the taint of vested interests on the promoters of fluoridation. The argument is that a number of industries potentially gain from fluoridation: aluminium companies that produce fluoride as a waste product; toothpaste companies that promote their product using fluoride; and sugary food manufacturers that benefit if tooth decay is seen as due to lack of fluoride rather than the presence of sugar. The antifluoridationists have discovered a number of cases going back to the 1940s where individuals linked to some companies in these categories have been involved in the promotion of fluoridation. If this connection can be established, the implication seems clear: fluoridation was introduced because of financial considerations, not primarily because of its alleged benefits.

With a bit of imagination, the charge of vested interests can be brought against just about anyone. In the case of occupational overuse injury, opponents of the medical explanation have charged that doctors have received generous fees from patients that they diagnose with this problem and also, in some cases, for testifying in court cases defending claims made by workers for compensation.

The critics of nuclear winter have charged that the proponents are in league with the peace movement. They note that leading figures promoting nuclear winter are sympathetic with peace movement goals, and that the peace movement has played an important role in publicising the nuclear winter findings. This is a claim of an ideological vested interest rather than a financial vested interest. (The money involved has been used to promote nuclear winter rather than enter the pockets of the scientists.)

Financial interests can be indirect, in which case they can be called career or professional interests. For example, many of the leading proponents of nuclear power are nuclear scientists and engineers. Because that is their field, their careers are likely to benefit as the nuclear industry expands: there will be more positions, more promotions, and more workers influenced by their stands.

If there had been no nuclear weapons or nuclear power stations, it is most unlikely that Edward Teller would have obtained a key position of influence in US science policy. This is a professional interest.

Then there is psychological interest. The people who back a cause frequently tie their reputations to it. Its success represents their personal success, and vice versa. As a result, they are reluctant to recognise any evidence or argument that questions their cause. Virtually all prominent supporters of scientific orthodoxy have a psychological interest in their stand.

The defenders of medical orthodoxy on the connection between smoking and cancer arguably have both a professional and psychological interest in their view. If the orthodox view were undermined, the medical establishment would look very foolish. Furthermore, prospects for professional advancement by studying and promoting the connection between lifestyle and health would be jeopardised. Finally, the reputation and self-image of leading researchers and spokespeople would take a terrible beating. So it is reasonable to claim that defenders of the orthodox view have a strong psychological commitment that prevents them admitting any possibility of being wrong.

The charge of psychological interest is an effective comeback whenever defenders of orthodoxy become impassioned or abusive of opponents. When proponents of fluoridation dismiss their opponents as cranks and nuts, a suitable rely is that this simply shows how unbalanced the proponents have become: they are so personally threatened by criticism that they cannot conduct a calm debate sticking to the facts.

One of the troubles with pointing to vested interests is that the other side might do the same. If interests can be attributed to anyone, they can be stuck on you too. The opponents of fluoridation have been linked to chiropractors, Christian Scientists and health food companies. Opponents of nuclear power have been said to be serving the interests of the middle class. Critics of nuclear winter are said to be serving the military establishment. The catalogue continues.

Two points are relevant here. First, only some claims about vested interests can be made to stick. If your interests in a cause are not obvious, or not considered objectionable by most people, you have little to fear by pointing to vested interests on the other side. But if your interests are conspicuous, this tactic is risky. This is most obvious in the case of the tobacco industry, which has an obvious financial interest in questioning the link between smoking and ill health. Pointing the finger at the other side would seem hypocritical. But, on the other hand, if everyone sees you as tainted already, perhaps it can't hurt.

The other thing is that those attacking the orthodox position generally have less credibility. In most cases their sincerity will be attacked anyway, so they might as well point to vested interests linked to the establishment position. If there is mud being thrown around by both sides, it is more likely to help the side starting out in the weaker position.

Of course it's not just a question of throwing mud, but making it stick. A solidly researched case is useful, showing what the interests are and why they are important.

One response to the charge of interests is to admit that they exist, and claim that they don't matter: "So what? My stand is based on science, and everyone has an interest in the truth." In many cases it's not hard to ridicule this response, using a few choice counterexamples. The Swiss company Grünenthal that produced morning sickness drug thalidomide claimed that all the tests showed its drug was safer than any other — even after it had received considerable data showing its dangers.

The defence based on truth is not always easy to knock down. One way to undercut it is to point to a conflict of interests. The defenders of orthodoxy may have an official commitment to welfare, scientific truth or whatever, but this may conflict with other interests. For example, the promoters of nuclear power have a conflict of interest if they stand to benefit financially from its expansion.

Some people will recoil from the tactic of alleging vested interests. After all, it's not really anything to do with the facts of the matter. This view overlooks the way in which the facts are influenced by the interests of those presenting them. Nuclear winter scientists are not going to publicly draw attention to the limitations of their expertise in nuclear targeting or arms control. Nor are most biologists going to spend time recounting the evidence that contradicts conventional evolutionary theory. Whenever an issue enters the public debate, the "facts" presented in the debate are carefully selected and packaged for maximum effect. Inconvenient facts are brushed aside, and errors, gaps, assumptions, manipulations, extrapolations and a whole range of operations on the facts are papered over. Sometimes the technical literature in the area is also purged of admissions of shortcomings, since opponents can take advantage of the smallest weaknesses.

In the face of this process, can one really expect truth to emerge? The pointing out of vested interests can actually help to promote the truth, since it makes people aware of the limitations of the facts, apparently glossy and pure, as they are tossed into the debate.

The key ethical choice is not whether to point to interests, but *how* one points to interests. The modest or polite method is to emphasise that interests may have influenced the presentation of facts by the other side, in an attempt to show what the proper considerations really are. The more aggressive method is to point to interests as entirely undermining the credibility of the experts.

The problem, of course, is that it is hard to use the first method without using the second one too. Since the experts typically claim to be repositories of unsullied truth, when you show the smirches on their truth, it rubs off on the experts themselves. But then again, that may be the whole point of the exercise.

Attack the character of individual experts

Dr William McBride is a Sydney doctor who in 1961 was one of the first to draw attention to the deformed babies born to mothers who took thalidomide while pregnant. McBride won a major award as a result, and set up a medical research foundation, called Foundation 41, to study the first 41 weeks of human life, from conception to just one week after birth. Over the years McBride took stands against some other drugs, citing research evidence showing effects on fetuses.

In December 1987, Dr Norman Swan of the Australian Broadcasting Corporation revealed evidence that McBride had altered figures in a scientific paper dealing with the effect of the chemical hyoscine (related to the morning sickness drug Bendectin). In other words, Swan accused McBride of scientific fraud. The pressure mounted on Foundation 41 to take action, and eventually an independent committee was set up to investigate. McBride was found guilty. He resigned from Foundation 41, his reputation in tatters. Investigative reporters delved into his past, revealing facts that put some of his earlier activities and achievements in a more dismal light.

Fraud by scientists is more common than generally realised. But it is seldom exposed. To some people, McBride's sins would seem small: he had changed a few figures, and reported results for some rabbits that had never existed. But fraud is considered an extremely serious allegation against a scientist, and even minor violations are treated gravely. Penalties can be severe.

The problem is that it is hard to prove allegations of fraud, especially against powerful scientists such as McBride. Usually only co-workers realise what is going on, and they often depend for their jobs and references on the good graces of their superior. In McBride's case, it took five years before the allegations were brought to public attention. The junior researchers who first confronted McBride and Foundation 41 with their concerns left their jobs. Foundation 41 at that time took no action against McBride, and did not even investigate seriously. It took the persistence of a crusading journalist to bring the matter to public attention before the scientific community took action.

The first message here is that reputation is absolutely vital to experts. Violations of proper behaviour can undermine a reputation even for things unrelated to the violation.

The second message is that it is very hard to prove allegations of serious misconduct. It is risky to make claims of fraud or other violations of proper behaviour without solid evidence.

Fraud is perhaps the most powerful weapon against a scholar, more than assault or robbery. This is because fraud is a violation of what is considered proper scholarly behaviour. To expose a fraud can undermine an entire research area.

The most famous case in recent decades involved Sir Cyril Burt, a British psychologist who did studies on identical twins showing that the effect of genetics on IQ was much greater than the effect of environment. Burt's work was highly influential in British educational policy, helping to justify the use of examinations to select students for different schooling streams.

It was only in the 1970s, after Burt died, that close attention by critics to his research findings showed that his figures on IQ and inheritance were too good to be true. Closer investigation showed that in at least some of his work, Burt had manufactured IQ scores in order to give the results he expected to obtain. He had also apparently added names of collaborators who had done no work to some of his papers, to give the impression that others were involved in the research.

Once Burt's fraud was revealed and publicised in the press, it was a severe blow to other researchers subscribing to Burt's orientation on inheritance and intelligence, even though they were not implicated in any fraud. Their orientation can easily be denigrated by referring to it as one that for many years relied heavily on the fraudulent work of Burt.

One special category of fraud is plagiarism, which essentially means using someone else's work without giving proper acknowledgement. Plagiarism is also more common than generally acknowledged. Some types of plagiarism can be exposed relatively easily, especially word-for-word copying, simply by comparing the original source with the copied version. A serious case of plagiarism, if exposed and pursued, can be enough to bring down a scholar's career. On the other hand, most plagiarists get away with it, since it is very risky for junior scholars to make allegations. Institutions usually side with the powerful.

Another transgression can be called exploitation. This occurs when senior researchers take credit for the work of subordinates. This not only is common but is standard policy in many research laboratories. One powerful and prestigious scientist may run a lab in which dozens of junior researchers, often mostly PhD students, do the day-to-day work. The senior scientist is typically included as a co-author of many or all the publications produced from the lab. So the senior scientist seems to be incredibly, indeed impossibly, productive. The junior researchers are being exploited: part of the credit for their work is claimed by their boss.

Prominent experts often fall into this category. They may not have done any research with their own hands for years, but simply supervised others, or just raised the funds for the lab. Experts in this situation can be challenged. Did they really do the work for all their publications? Do they really understand the technical details?

This sort of misallocation of credit is standard policy in most government bodies. Credit for work done by junior staff is taken by senior staff. Statements by top officials are drafted by underlings, as are the speeches of most politicians.

One of the most prominent experts certifying the smoking-cancer connection is the Surgeon-General of the United States. When Surgeon-General Koop speaks, he speaks with authority. But does he speak with expert knowledge? Almost certainly he has not even read many of the research papers that back up his stand, much less done any of the work himself. At best he has been briefed by the experts in his department; at worst he has simply signed statements put on his desk.

Does it really help to expose the misallocation of credit and the exploitation of junior staff that is common in science? The answer is yes, because it undermines the prestige of the top figures who are the most common spokespeople for official policy. Let the real experts, those unrecognised junior people, speak for themselves. Of course, the junior researchers do not have the same status. Their research briefs are narrower, and can be criticised as not relevant to the wider public issues of concern. They are unlikely to be experienced in public debate, and are more likely to make concessions or damaging admissions. Most of all, they are unlikely to be as unified in support of orthodoxy as the few official spokespeople at the top.

For these reasons, it is worthwhile to expose fraud, plagiarism, corruption, exploitation and any other violations of proper procedure that can be attributed to the most prestigious figures. But beware! It is not easy to make allegations of this sort stick. Proceed with care.

The proponents of nuclear winter have been accused of several violations of proper scholarly behaviour. One of the critics, Russell Seitz, alleged that the important article by Carl Sagan and his colleagues, published in the prestigious journal *Science* in December 1983, was helped through the refereeing process by being given only referees chosen by the authors. Seitz also complained about pre-publication promotion of the nuclear winter theory, fostered by Sagan and company, which is contrary to the traditional view that scientists should present their work in scholarly venues (journals and conferences) before any reporting in the mass media occurs. Seitz's implication was that nuclear winter was being touted more in the manner of a public relations exercise than a carefully considered scientific discussion.

There is other information about experts that can be damaging when revealed. If experts have criminal records, have spent time in mental institutions, are drug addicts, batter their spouses and children, or consort with criminals, making this known can help undermine their credibility. Needless to say, you must be very cautious when making such claims about a person's behaviour.

Remember that information of this nature probably does not actually have anything to do with a person's expertise, in a strictly logical sense. Just because a person has been convicted of shoplifting or embezzlement should not, in principle, affect their expertise on missile accuracy or chemical reactions. An expert who responds in this fashion is making a sound point. What you can argue, or suggest, is that a person's overall credibility as a person must be considered. Anything an expert does can affect this.

Another risk in bringing up the sordid side of an expert's life is actually generating sympathy for the expert. For example, the late Sir Ernest Titterton, one of Australia's most prominent nuclear proponents, had worked with radioactive materials for many years and was known to have children with genetic defects. This information suggests an insight into the psychology behind Titterton's declarations that low-level ionising radiation is not harmful: a refusal to accept the possibility of being responsible for harming his children. But exposing such information could have backfired. People might have felt sorry for the children and for Titterton. Also, people might have been repelled by the act of bringing such personal information to the public eye simply to win some points in a public controversy. Finally, there is no proof that the genetic defects were actually due to radiation exposure. (This information was known in the Australian antinuclear movement for years, but never used publicly.)

Attacking the character of an expert is playing with fire. It can totally destroy an opponent, but it can also get out of hand. The most appropriate use of this technique is when the flaws are highly relevant to a person's expertise, such as scientific fraud. In other cases, great care is required.

Summary

If the experts are lined up on the other side, you may have little chance of making headway by simply dealing with the facts and the assumptions underlying the facts. As long as the other side has the credibility of credentials, endorsement by official bodies and respect from other professionals, your facts and counterassumptions will bounce off the virtually impenetrable wall of authority. Therefore you may choose to attack the experts themselves.

Exposing their mistakes, their false predictions and their exaggerations is always effective. No expert is right all the time. If you can show that the expert has been wrong one time, then why not on this issue this time?

Experts sometimes change their views to suit the tenor of the times. Most people don't notice since they don't follow the issues for many years and check up on what the experts are saying. A bit of investigative work can show inconsistencies that can discredit the expert.

Don't be awed by credentials: attack them, or at least attack their relevance to the issue at hand. Why should a person with a PhD on one plant species be considered an expert on whether pesticides are economically beneficial or socially acceptable? Most expert credentials are not directly relevant. They certify the person as a expert in an extremely specialised area. Be sure to ram this point home.

You can almost always show that an expert has some sort of vested interest in their position. Sometimes they stand to gain financially or in future career prospects. Sometimes they have ideological links to a social movement of whatever complexion. And in almost every case they develop a psychological interest in the position they have so vocally defended.

Exposing interests is quite legitimate. Facts and arguments do not exist in a vacuum. The choice of facts, the development of arguments, and the money and human enthusiasm used to promote them all depend on interests. This can be the vested interests of a nuclear industry, the ideological interest of a peace movement, or the psychological interest of an evolutionary biologist.

Finally, it is possible to attack an expert by exposing "character flaws." This can be scientific fraud, violations of proper scholarly behaviour in seeking media coverage, or a criminal record.

If you think attacking the expert is not proper, just stick to facts and assumptions. If you are effective, it probably won't be long before the experts start to attack you personally, rather than just your facts and assumptions. Counterexperts are prime targets for attack by establishment experts.

Investigating experts

In order to mount an attack against experts, you need to gather information. You need copies of articles written by the experts, plus further information about them. There are many sources of information.

You can start by asking the expert directly for copies of articles and talks. Many will be happy to oblige. If they are willing to send a full list of their publications, much of your work is over. Just arrange to get copies at the nearest big library, or through interlibrary loans.

But some experts will not help this much, and in any case they will usually only send their technical writings. Furthermore, you also will want, perhaps most of all, records of their stands in public debate: talks, articles in popular journals, and letters to the editor.

Start at a good library and ask the librarian about indexes to the relevant "literature." For example, there is the *Science Citation Index*, the *Social Sciences Citation Index*, the *Public Affairs Information Service* and many others. These will give references to articles published. Once you get some articles, look through them carefully and obtain others that are cited there.

If you know the expert has been writing articles or letters to particular newspapers, you may want to spend the time to go through back issues — but it does take time. Some newspapers and libraries keep cuttings files on prominent figures, though these are usually incomplete. So do various organisations, such as some government bodies, politicians, and environmental groups. Try every possibility and combine the results.

You can develop your own file by reading the local press. Also, you can subscribe to press cuttings agencies that, for a fee, send you cuttings of articles on particular topics or individuals, from around the country.

Finally, you should contact leading critics to find out what information and knowledge they have about the experts. Often they have copies of material unavailable elsewhere.

Collecting copies of articles and speeches is a solid basis for studying the views of experts and being able to expose mistakes, exaggerations and inconsistencies. If the expert says one thing to one audience and another thing to another audience, you will have good evidence. You have the added advantage of seeing what arguments the expert uses regularly, which facts and counterarguments are dodged, and whether the expert "recycles" text from one article to another (self-plagiarism).

To obtain information about the expert's life, you can start with sources such as *Who's Who*. Some of the articles you have collected will give biographical information, if only where the person works. You can use this as a basis for probing further. For example, if the expert works at a university, you can inspect annual reports to find lists of publications, when promotions occurred, and perhaps some public relations material about activities.

Using the techniques above, you can collect an extensive file on the expert. Then what? You may prefer to use the information selectively, in debates or letters to newspapers. Alternatively, you could compile a dossier or article for distribution. Before you distribute it widely, you need to make a crucial choice: should you send it to the expert for comment? If you do, you reveal your information and allow the expert to prepare a defence. But there are advantages in sending your draft dossier or article to the expert. If you have made any mistakes, the onus is on the expert to correct them. Otherwise, you can point out that the expert didn't offer any corrections. Also, you are setting a

DISCREDIT EXPERTS

standard for behaviour. How would you like to have a dossier about *you* circulated widely, without being consulted? Finally, just sending a dossier to an expert can make all the experts on that side more cautious: they will know that someone is watching them closely. If they adopt a lower profile, that may help your cause.

4
Discredit Expertise

"Expert: a person who avoids small errors as he sweeps on to the grand fallacy." — Benjamin Stolberg

At times it may be advantageous to expand the attack. Instead of just attacking certain experts, you can target expertise itself. This is a major step which, if taken seriously, has wide-ranging implications. If expertise is undermined, then issues must be dealt with by non-experts. It might seem tempting for the freedom-to-smoke supporters to deny that reliable expertise is possible on the issue of smoking and health or on smoking and civil liberties. But this would not necessarily stop the antismoking campaigners, most of whom are not experts anyway. They could easily continue just on the basis that smoking is unpleasant and objectionable.

Let me first describe some of the ways to attack expertise, and then return to the occasions when this approach is likely to be most useful.

Expose social processes in knowledge creation

The traditional picture of science shows the scientist working away in the laboratory, eventually making a brilliant discovery (or, more likely, a minor advance). The new knowledge springs directly from the head of the great thinker, is written down and immortalised in a publication. Einstein's discovery of the special theory of relativity, and its publication in 1905, is a well-known model of how this is supposed to work. Einstein became famous after his general theory of relativity was verified by observations in 1919 of the planet Mercury.

Karl Popper, a famous philosopher of science, was not satisfied with verifications. He demanded that attempts be made to try to show that any particular scientific hypothesis was wrong. In other words, attempts should be made to falsify, rather than verify, scientific claims. If it was impossible in principle to be able to show that an idea was false, then Popper said it wasn't science. On this basis, Popper attacked psychoanalysis and Marxism for being unscientific. Neither could be proved wrong, since neither made any prediction that could be falsified. Naturally, psychoanalysts and Marxists disagreed.

Popper allowed for scientists, as individuals, to be wrong, foolish or crazy. That was no problem, for a scientist's conjecture was not scientific knowledge. What was required was a test of the claim by others: an attempt at refutation. Science, Popper said, is a process of conjectures and refutations. Therefore scientific knowledge results from a social process. The result, though, said Popper, is not tainted by its origins in ordinary mortals.

This traditional view of science is well liked by experts. The experts, after all, rest their reputation on the image of their knowledge as being above and beyond the foibles of individuals. Few experts know much about the philosophy of science, and only some are familiar with the doctrines of positivism and falsificationism. But that doesn't matter. Their critics are seldom better versed. But they ought to be, because some of the new views of scientific knowledge are more of a threat to the usual pretensions of experts.

The new approach is usually traced to Thomas Kuhn and his book *The Structure of Scientific Revolutions*, first published in 1962 (though others had similar ideas earlier). Kuhn said that science normally progresses by small increments, as scientists carry out research under the prevailing set of ideas and methods. This prevailing set of ideas and methods is called a paradigm.

Occasionally the whole set of prevailing ideas in a field is challenged and overturned. This is called a scientific revolution. A new paradigm is set up. In physics, the classical paradigm was Isaac Newton's laws of motion. Then Einstein came along and said that at high speeds, funny things start to happen: time slows down, objects get longer, and so forth. Relativity was the basis for a scientific revolution in the field of physics.

Kuhn was concerned with traditional natural science, but others have taken up his ideas and applied them to all sorts of fields: psychology, economics, medicine . . . even fluoridation. Whenever there is a major clash between different ways of analysing the world, this can be called a paradigm conflict.

The Kuhnian picture is especially useful for critics of orthodoxy. Rather than the standard view being "the truth," it becomes just the current paradigm. The orthodox experts are "paradigm-bound," that is, they are stuck in their way of doing things and can't fully comprehend or assess alternative ways of conceiving the world.

For example, supporters of fluoridation can be said to be caught in the "fluoridation paradigm." They are tied to the view that fluoridation of water supplies is natural, beneficial and harmless. Because all their research takes place within this viewpoint, they do not undertake studies that might cast doubt on their basic assumptions. The antifluoridation paradigm, by contrast, looks for the problems with fluoridation and, not surprisingly, finds them.

The concepts of paradigm and revolution have several advantages for those challenging the experts. They take the orthodoxy out of the category of everlasting truth and into the category of a provisional way of doing things. They also cast orthodoxy into the role of the establishment, which ultimately can be overthrown by a revolution. Many people are suspicious of establishments and sympathetic to challengers, who are the underdogs. It's better to be an underdog making a challenge to a provisional way of doing things than to be a crank who disputes an undeniable truth.

The idea of paradigms doesn't undermine expertise, since any paradigm has its experts. The next step goes beyond Kuhn. The key question is, why do paradigms develop the way they do? In other words, why do scientists prefer one framework of ideas over another?

All sorts of factors can enter in here. Money, for example. Chemical companies pay researchers to study the effects of their pesticides. This provides an incentive to use the "pesticide paradigm" in which the only solution to the problem of pests is to kill them with chemicals.

Belief systems can shape paradigms. Nuclear winter researchers are sensitive to the vulnerability of the environment to human impacts, arguably because they are sympathetic to the environmental and peace movements. What they think of, look for and build into their models then shows up in their results. This is one facet of the "nuclear winter paradigm."

It isn't hard to see that every factor that is used to discredit facts, arguments and individual experts can be used to discredit a whole body of knowledge. Paradigms can be shaped by money, possible jobs, bureaucratic vested interests, professional status, ideology and a host of other factors.

Sociologists have also looked at the day-to-day activities of scientists. What have they found? Essentially, scientists are involved all the time in making value judgements and in persuading and being persuaded by other scientists and by outsiders. This applies to every detail, including deciding what counts as a fact. Nature does not pop into the lab and point a finger at some evidence, saying "that's a fact." The scientists must interpret what they search for and find, and there is *always* plenty of room for competing interpretations. For something to become a fact, other scientists must be convinced. That means persuading them that a particular way of seeing things is appropriate.

The verification of facts and testing of theories by other scientists always involves elements of persuasion. Other scientists have to be convinced, in one way or another, that it is worth verifying a claim by another scientist. In many cases they do not bother, since they think they know in advance what they will find — especially when the other scientist is considered a crank.

The point of all this is that the process of scientific inquiry is shot through with personal factors which may be

influenced by the wider politics of the issue. In the case of
fluoridation, the opponents argue that proper checking of
claims of harm from fluoride has not been made. If so, this
could partly be because antifluoridationists have little
scientific credibility, or because little money is made
possible for research potentially critical of fluoridation, or
because scientists who do research critical of fluoridation
have difficulties in their careers. In each case, a political
factor, whether credibility, money or career prospects, can
influence the development and assessment of scientific
facts.

Strictly speaking, just because scientific facts are
negotiated on a day-to-day basis, with this process being
influenced by the personal desires of the scientists as well
as wider political factors, does not mean that the knowledge
generated is useless. In many cases it works very well, in
the sense that other scientists come up with the same results,
or use it to develop further knowledge.

Yes, the knowledge generated in this fashion can be
useful, but who is it useful for? The knowledge produced
relevant to fluoridation, in a context in which credibility,
money and careers favour fluoridation, is likely to be
useful for promoting fluoridation. Surprise! If antifluoridationists had more credibility, dispersed the funds and
provided the jobs, the findings would more likely favour
them, as is the case in some countries such as India.

One way of looking at this is to say, a profluoridation
research environment is likely to produce results which
are "selectively useful" to supporters of fluoridation. Some
of the results can still be used by antifluoridationists
(which is what happens), but most results are easier to use to
promote fluoridation.

The sociologists who study scientific knowledge say that
there is no way to escape this situation. Scientific
knowledge is always created by scientists who are influenced by their context. They are concerned about their
status, salaries and prestige; they are influenced by the
ideas of dominant groups in society; they are influenced by
the professional hierarchies and bureaucratic structures in

which they work; and they are influenced by the availability of research funding and laboratory facilities to carry out certain types of research. Occasionally some of them are censored or sacked for doing the wrong thing, which has happened to some antinuclear and antifluoridation scientists.

As I said, strictly speaking this should not discredit expertise, but simply make clear the context in which it operates. In practice, describing the social processes and political environment of science *does* serve to discredit it. This is because science has been sold to the public as objective knowledge that is untainted by social factors. As long as this mythical picture is taught to unsuspecting students and portrayed in the media, a social analysis of the actual practice of science serves to undermine expertise.

Deny the relevance of expertise

It is a short step from denying the relevance of the credentials of particular experts to denying the relevance of expertise generally. The easiest way to do this is to point out the aspects of an issue that require value judgements and public decision-making.

Experts on the construction of nuclear power plants are not specially qualified to pass judgement on whether nuclear power is an appropriate part of energy policy.

Experts on evolutionary biology are not specially qualified to pass judgement on educational policy, which might validly include exposure to a range of viewpoints, including creationism.

Experts on smoking and health are not specially qualified to pass judgement on whether smoking should be advertised or prohibited in public places.

When it comes to a detailed assessment of expertise, it is amazing how little relevance most of it has to the issues that matter to most people. A nuclear scientist might be an expert on a particular type of nuclear reaction. A biologist might be an expert on the habitat of a particular species. A medical researcher might be an expert on the development of a particular type of cancer in rats. Clearly, the narrow expertise of such researchers has little relevance to issues of public policy.

It is best to be prepared for "trust the expert" arguments with a few counterexamples. Argument: "I wouldn't want a nonexpert flying any aeroplane in which I was a passenger." Answer: "Let the aeroplane pilots do their job, but don't let them tell us whether to travel by car, train, ship or plane."

Argument: "I want an expert to fix my broken arm." Answer: "It's fine for doctors to fix broken arms, but they are not suitable experts for deciding public policy on smoking, drugs and health insurance systems."

It is even easier to argue against expertise that serves vested interests. Obviously, we should not trust the manufacturer of breakfast cereals to tell us what to eat for breakfast, nor should we trust the shoe manufacturer to tell us what clothes to wear. Don't ask the barber whether you need a haircut. Don't ask the encyclopaedia merchant whether you need an encyclopaedia.

These analogies may sound silly, and that is exactly what the argument about trusting experts is. Most experts are remarkably narrow in training and experience. They are precisely the wrong people to be providing general direction for society.

By all means, let us consult the experts, but don't let them tell us what to do. As the saying goes, experts should be "on tap but not on top."

Although most experts are narrow, there are some whose expertise is broader and seems more relevant. For example, a medical epidemiologist may study statistical patterns of health and disease as a function of diet, occupation or habits such as smoking. This seems more relevant to public policy. But remember, public policy is just that: it should result from informed participation and informed consent. No expert can pretend to speak for the values of others. Those values are expressed by their beliefs and actions.

Expertise is overrated. In most issues of importance to most people, the relevance of expertise is remarkably small. What the experts try to do is convince people that their credentials and their knowledge entitles them to pass judgement on anything to do with their area of expertise. (Often it is powerful groups such as governments and corporations that use the experts to promote their policies.) What the critics have to do is expose this illegitimate expansion of the domain of the experts, and demand that experts restrict themselves to their narrow territories.

Expose interests

Experts collectively have a vested interest in expertise becoming a basis for status, power and wealth. This fact provides a basis for attacking expertise generally, at least the way it is organised and used in present-day society.

There are even theories about the vested interests of experts. The simplest version is contained in the term technocracy, rule by the experts. Nothing like this seems in the offing. Prime Minister Margaret Thatcher may have been a chemist, but her scientific expertise was not the basis for her rise to power.

Rather than direct rule by experts, a better explanation of the rise of experts to power is through changes in the way traditional systems of power operate. Bureaucracies, both public and private, are the prime example. Within these systems based on chains of command and routine specialised work, knowledge is a crucial commodity. Information is sent up the hierarchy and instructions are sent down. Those people with claims to special knowledge can claim greater power, especially those at the top who have greater access to inside information.

In the debate over nuclear power, nuclear expertise has largely been tied up with bureaucracies, such as the US Atomic Energy Commission and its successors. Individual nuclear experts would not have much power as isolated individuals. But when speaking as representatives of a large organisation, or when speaking as individuals backed up by the bureaucracy, their expertise has much more sway.

Similarly, promotion of fluoridation has been founded on the backing of the US Public Health Service as well as the American Dental Association. Pro-fluoridation experts derive much power from these connections.

The expansion of the role of expertise in large organisations can be attacked as the rise of a New Class, or Intellectual Class. Some critics of Soviet-style societies use this sort of analysis. Socialist revolutions destroyed the power of capitalists. Who benefited? Intellectuals. To begin, most leaders of revolutionary parties, such as Marx, Lenin, Trotsky and Mao, have been intellectuals. (This pattern continues today — visit any vanguard left-wing party.) But more importantly, the introduction of state socialism means a massive expansion in government employment. Hoards of bureaucrats are required to run the country. Who gets these jobs? Intellectuals — people with credentials and specialist knowledge.

What happens quickly in a socialist revolution is happening more gradually in so-called capitalist societies. The management of work and life, from government bodies to large corporations, becomes ever more important. People rise to power beginning as credentialed intellectuals: lawyers become politicians; engineers become corporation presidents; economists become government bureaucrats.

These different groups have several things in common. They defend formal training and credentials as essential to gain entry into occupations. They demand that discussions take place in "reasoned" terms, the terms of intellectual debate. Moral indignation and principled stands are set aside in favour of neutral styles of discourse, using tools such as cost-benefit analysis. Finally, and most important, they promote a type of society in which specialist knowledge, when linked to power, is seen as legitimate and worthy of great social rewards.

In other words, experts collectively are usurping power that should legitimately be in the hands of people in a democratic society. For the rising New Class, the only tolerable form of democracy is one with representatives who are suitably responsive to the experts. For experts with access to power, populism is dangerous.

This critique of expertise is not really an attack on expert knowledge but rather an attack on expert knowledge allied to powerful institutions. The attack should focus on the vested interests involved. The antismoking medicrats are linked to government agencies that are exerting ever more power over our daily lives. The pronuclear experts are linked to government and industry bodies. The profluoridation experts are linked to government bodies and dental associations. The proevolution lobbyists are linked to government bodies that exert powerful control over school curricula.

Summary

Attacking expertise is an ambitious enterprise. A large fraction of people in powerful or high status positions are or were experts of one kind or another: journalists, lawyers, doctors, government bureaucrats, corporate engineers or managers. Yet at times it may be to your advantage to point out biases and flaws underlying expertise as a whole.

Recent studies of science have undermined the view that scientific knowledge has some unique path to truth. Social and political factors enter into the development of scientific ideas and into day-to-day scientific research activity. It is reasonable to argue that no part of science is neutral: it will always be tailored to be more useful to some groups than others. Scientists will claim otherwise, but then they have been taught a convenient myth.

Another way to attack claims based on expertise is to deny that expertise is relevant. Most experts are very narrowly trained, while most issues of social significance involve all sorts of issues in which the experts have no special brief, such as assessments of justice or the acceptability of risks. Expertise is usually not all that relevant. Experts claim otherwise, but then it is in their interest to do so.

Experts are part of the New Class or Intellectual Class. These are names for a roughly defined group of people who use claims about knowledge to advance their status, power and wealth. This includes government bureaucrats and corporate managers, among others. Members of the New Class prefer to define issues as technical ones that they are better qualified to deal with. In making such claims they are serving their own collective interests, and acting against a more populist, democratic method of dealing with social issues.

Tips on dealing with the experts

The methods of attack outlined in this booklet provide some guidance in deciding what weak points in the expert's case you want to target. But in an actual confrontation with an expert —in the letters column of a newspaper, in a formal face-to-face debate, or via media releases — there are many practical skills that are important. You can learn these skills by observing debates and joining them yourself.

There is no single best action that applies to all circumstances. What is best for you depends on what you want to achieve, the individuals and organisations you are up against, what has happened already, and various chance and unknowable factors.

So, instead of detailed advice, here are a series of tips. You can best use them as a checklist. When planning your next move, look through them and decide whether any particular tip is relevant. Often a tip will not be relevant, and sometimes you will decide a tip is the wrong advice for your circumstances. The tips are to remind you of possibilities. The most frustrating mistake is to forget something that should have been obvious! As you gain experience, you can create your own personal list of tips.

• Plan your strategy.

• Always check and doublecheck your facts.

• Before sending off a letter or article, ask at least one person you trust to read it through for flaws.

• Practice your speaking and writing skills regularly.

• Avoid arguing on terms set by the experts, especially on technical issues. Emphasise the issues you think are important.

• When an expert makes a mistake or reveals an embarrassing assumption, ram this home again and again.

- Carefully study what the experts say and write; know who and what you are up against.
- Be prepared for dirty tricks from the other side.
- Keep cool and don't act in haste. The idea is to open up the issue, not just let off steam.
- Keep pressure on the experts. Some of them will do something foolish in anger.
- It is more important to persuade sympathetic and neutral people than to try to win over those on the other side.
- Help others to join in on your side rather than doing everything yourself.
- Imagine yourself in the position of a sincere expert on the other side. How would you react? Use the insights from this sort of thinking to develop your tactics.
- Consider using humour in your campaign. Few experts can tolerate being laughed at.
- Set some easy short-term objectives to build confidence.
- Carefully study your defeats.
- Celebrate your successes — but don't become complacent.
- Be prepared for a long struggle.

References

There are untold writings dealing with experts and expertise in one fashion or another. I list here only a few that I have found useful in understanding and challenging experts.

General references

Stanislav Andreski, *Social sciences as sorcery* (New York: St. Martin's Press, 1973). A demolition of social science.

Phillip M. Boffey, *The brain bank of America: an inquiry into the politics of science* (New York: McGraw-Hill, 1975). Case studies of experts and politics.

Christopher Cerf and Victor Navasky, *The experts speak: the definitive compendium of authoritative misinformation* (New York: Pantheon, 1984). A delightful collection of quotations by experts.

David Elliott and Ruth Elliott, *The control of technology* (London: Wykeham, 1976). An excellent discussion of the relation of experts to systems of power.

Ivan Illich, *The right to useful unemployment and its professional enemies* (London: Marion Boyars, 1978). An attack on industrial society serviced by expert professionals.

Harold J. Laski, *The limitations of the expert* (London: Fabian Society, Fabian Tract No. 235, 1931). An incisive 12-page critique.

Jethro K. Lieberman, *The tyranny of the experts: how professionals are closing the open society* (New York: Walker, 1970).

Dorothy Nelkin, "The political impact of technical expertise," *Social Studies of Science*, Vol. 5, 1975, pp. 35-54.

Joel Primack and Frank von Hippel, *Advice and dissent: scientists in the political arena* (New York: Basic Books, 1974). Case studies of expertise and politics.

Michael Young, *The rise of the meritocracy 1870-2033: an essay on education and equality* (London: Thames and Hudson, 1958). An amusing satire highlighting the link between formal education and privilege.

Fluoridation, nuclear power, etc.

Mark Diesendorf, "Science under social and political pressures", in David Oldroyd (ed.), *Science and ethics* (Sydney: University of New South Wales Press, 1982), pp. 48-73. An expose of assumptions in the pronuclear and profluoridation positions.

Leon J. Kamin, *The science and politics of I.Q.* (Potomac, Maryland: Lawrence Erlbaum Associates, 1974). A devastating attack on hereditarians, including exposure of suspicious features of Sir Cyril Burt's data.

* Brian Martin, *Nuclear knights* (Canberra: Rupert Public Interest Movement, 1980). A critique of nuclear experts Sir Ernest Titterton and Sir Philip Baxter.

* Brian Martin, "The naked experts," *Ecologist*, Vol. 12, No. 4, July/August 1982, pp. 149-157. A critique of nuclear expert Leslie Kemeny.

* Brian Martin, "Nuclear winter: science and politics," *Science and Public Policy*, Vol. 15, No. 5, October 1988, pp. 321-334. The uses of expertise in the nuclear winter controversy.

Allan Mazur, "Disputes between experts," *Minerva*, Vol. 11, No. 2, April 1973, pp. 243-262. The use of arguments by pro and anti experts on nuclear power and fluoridation.

Bill Nicol, *McBride: behind the myth* (Sydney: Australian Broadcasting Corporation, 1989). Expose of medical researcher William McBride.

---

* Available from the author at Department of Science and Technology Studies, University of Wollongong, PO Box 1144, Wollongong NSW 2500, Australia.

Charles Schwartz, "The corporate connection", *Bulletin of the Atomic Scientists*, Vol. 31, No. 8, October 1975, pp. 15-19.

The intellectual class

Michael Bakunin, *Bakunin on anarchy* (Sam Dolgoff, ed.) (New York: Vintage, 1971).

B. Bruce-Briggs (ed.), *The new class?* (New Brunswick, NJ: Trans-Action Books, 1979).

Robert J. Brym, *Intellectuals and politics* (London: Allen and Unwin, 1980). On the political orientation of intellectuals.

Eva Etzioni-Halevy, *The knowledge elite and the failure of prophecy* (London: Allen and Unwin, 1985). Argues that the intellectuals have not helped society.

Alvin W. Gouldner, *The future of intellectuals and the rise of the New Class* (London: Macmillan, 1979).

George Konrád and Ivan Szelényi, *The intellectuals on the road to class power* (Brighton: Harvester, 1979). Intellectuals are becoming the new ruling class, especially under state socialism.

Max Nomad, *Rebels and renegades* (New York: Macmillan, 1932). Unflattering portraits of revolutionary leaders and their role as intellectuals.

The nature of science

Randall Albury, *The politics of objectivity* (Geelong: Deakin University Press, 1983).

Thomas S. Kuhn, *The structure of scientific revolutions* (Chicago: University of Chicago Press, second edition, 1970).

BRIAN MARTIN
Uprooting War

FREEDOM PRESS

298 PAGES   ISBN 0 900384 26 3   £5